A Guide To
The Mines Of The Cripple Creek District

by

Bill Munn

Published by

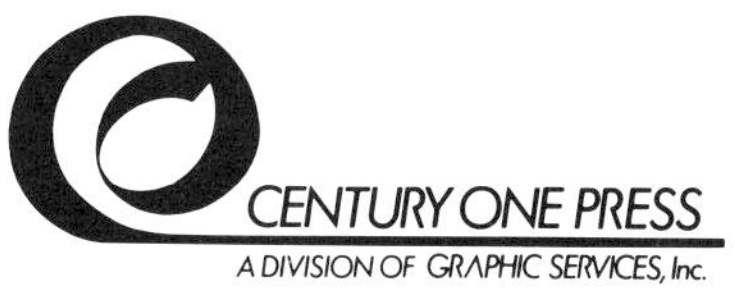

2325 E. Platte Avenue
Colorado Springs, CO. 80909

Library of Congress Catalog Card No. 84-70586
ISBN No. 0-937080-15-2

Cover photo: the Vindicator Mine

On the hills above the towns of the Cripple Creek District in Colorado, sometimes hidden among the aspen and evergreens, lie the remains of some of the most famous gold mines in the West. Each year, vandals, the weight of heavy snowfalls, and other elements destroy more of the little that is left of this rich western heritage. This publication is an attempt to preserve by photography some of this romantic legacy.

In the early years of the Cripple Creek District, around the turn of the century, it was difficult to separate fact from fiction, profit from promotion, so great was the amount of information available. As the productivity of the District waned and mining operations and operators struggled to stay afloat, the ancestry and development of the mines became much more difficult to determine. Quite a bit of effort has been made to identify each mine accurately.

Today, no trespassing warnings are strictly enforced, and for good reason. Not only have careless and inconsiderate individuals destroyed much of what remained of old townsites and mine buildings, but unknowing explorers risk fatal plunges to the bottom of water or gas-filled shafts. Caving ground and abandoned but still lethal explosives also await the unwary. Also, several active mining operations are currently blasting and chemically treating ore in the Cripple Creek District. This publication is designed to give you a firsthand look at the mines with none of the inherent risks. Beware!

The photographs, except for a few early day pictures, are from the author's collection. Available information was checked and rechecked to assure that each mine was correctly identified.

The author wishes to thank Mrs. Olive Hutchins and Mrs. Dorothy Henry, librarians, Lamar, Colorado; Marcia Golinko and Cavan Daly McGrew, runners, Colorado Springs; Mrs. Florine Parsons, Librarian, Enid, Oklahoma; Mary M. Davis, local history librarian, Pikes Peak Regional Library, Colorado Springs, Colorado; John Strauss and Joe Vanderwalker of Victor; Joan Drumright, Enid, Oklahoma, typist; and my wife, Kay.

Bill Munn
Enid, Oklahoma

January, 1985

Contents . . . Maps & Photographs

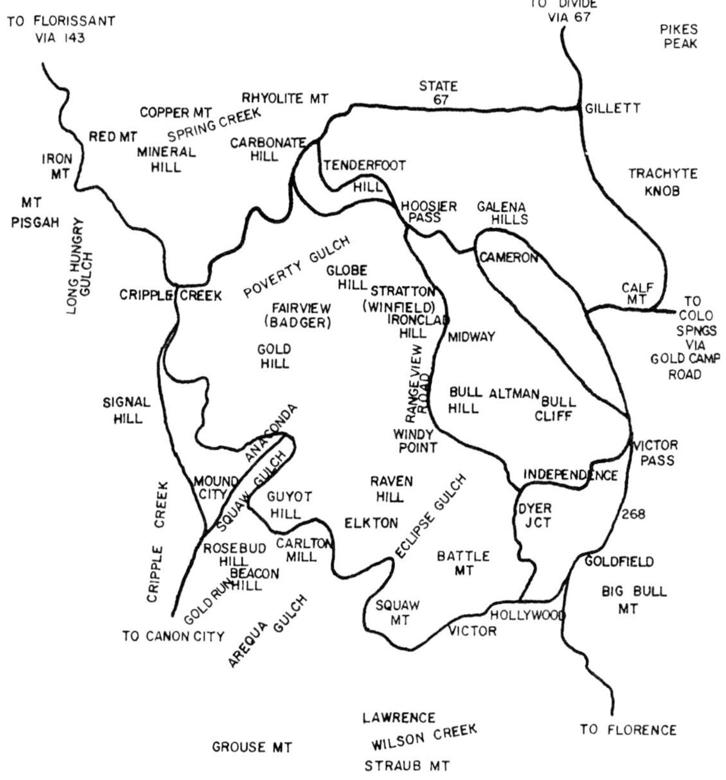

TOPOGRAPHY AND SETTLEMENTS
IN THE
CRIPPLE CREEK GOLD MINING DISTRICT
NOT TO SCALE

Introduction

This book is intended to guide the interested observer through the Cripple Creek Mining District in a logical progression from one area to another. The tour is on roads that are, for the most part, passable for most family passenger cars. Keep in mind that most mines can be observed *from these roads* and it is not only unnecessary but dangerous to leave the immediate area of the roadbed. The mines are divided into the following groups:

Poverty Gulch is the cradle of the Cripple Creek Mining District. Some of these mines can be seen from Cripple Creek but the majority are better viewed from the road leading east from the Mollie Kathleen Mine. They include the Abe Lincoln, Chicago and Cripple Creek Tunnel, Proper, C.O.D., Gold King, Mollie Kathleen, Black Diamond Tunnel, Hoosier, Friday, and Sangre de Cristo.

The mines of the North Slope of Bull Hill and Bull Cliff may be best viewed looking south from the Cameron townsite on the Gold Camp Road. Cameron was the site of Pinnacle Park, an amusement center, and the Forest Queen, WPH, Jerry Johnson, Damon, Cameron, Mitchell, Pinnacle, Blanche, Morning Star, Empire Lee, Buena Vista, Isabella, Smuggler Tunnel, Victor, School Section and Sunshine mines.

Midway was a favorite watering hole for miners coming from shifts on the Solomon, Patti Rosa, Wild Horse, Gleason, and Ironclad mines. The Golden Circle Railroad and the Cripple Creek District Railway or High Line met here between Ironclad and Bull Hills.

Windy Point offers one of the best panoramic views of the Cripple Creek District and the Continental Divide to the west. In this area are located the Amanda, Happy Year, Blue Flag, Rattler, Sheriff, Bogart-Twin Sisters, Nightingale, War Eagle, Ramona, and Favorite mines.

Also from this site can be seen many of the mines of Eclipse Gulch. They include the Maggie, Gold Sovereign, Dante, Trilby, Cresson, Trail, Rose Nicol, Ocean Wave, Coriolanus, May B, Sunset Tunnel, Eclipse, Carbonate Queen, Ada Belle, Moose, Bertha B., Joe Dandy, Ida May, and Frank R. Some of these Eclipse Gulch mines can also be seen from the Los Angeles and Eclipse mine sites.

The Six Points area is located near a saddle between Bull Hill and Battle Mountain. Mines located here are the American Eagles, Logan, Orpha May, Rubie, Sacramento, Blue Bird, Dexter, Los Angeles, Colorado City, Last Dollar, Modoc, Clyde, Wisconsin, Portland No. 3, and Rigi.

From the site of the Last Dollar Mine to the site of Independence town, the mines at Altman and Independence can be observed. Altman mines include the Deadwood, South Burns, Zenobia, Pharmacist, Burns, and Pinto.

Southeast, down the slope from Altman lies Independence, site of the Trachyte, Shurtloff, Delmonico, Atlanta, Findley, Hull City, Glorietta, Longfellow, and Sitting Bull. This is the scene of Harry Orchard's depot dynamiting, and traces of railroad spurs still cross the area.

Just north of the present town of Goldfield lie the Vindicator No. 2, Vindicator, Lillie, Anna J, Golden Cycle, Theresa, and Gold Knob mines.

Above the town of Victor lies the most productive ground in the entire District. Located here on Battle and Squaw mountains are the Santa Rita, St. Patrick, Gold Coin, Mary Cashen, Nellie V., Ajax, Granite, Lost Anna, Portland No. 2, Portland No. 1, Independence No. 2, Independence No. 1, Strong, and Jolly Tar.

The hill immediately southeast of the Carlton Mill offers a good view of mines on the south slope of Raven Hill and on the east slope of Beacon Hill. The mines called Elkton and Thompson are on the south flank of Raven Hill near Colorado State Highway 67. To the west the Gold Dollar, Prince Albert, and Beacon cling to the steep east side of Beacon Hill.

Besides the rich El Paso, several equally high-grade mines are located between Beacon Hill and Rosebud Hill. These include the Nichols, Maid of Orleans, CK & N, Little May, Old Gold, Henry Adney, Mary Nevin, and Hiawatha.

Little is left of the small town of Anaconda in Squaw Gulch. In the area identified by the huge stockwork dump of the Mary McKinney Mine, the Anaconda, Virginia M., Howard, LeClair, Raven Tunnel, Dolly Varden, Morning Glory, Doctor Jack Pot, Ingham, Wedge, Ophir, Katinka, Hobo, Rose Maud and Caledonia dot the surrounding landscape.

Down Squaw Gulch to the southwest of Anaconda, near Mound City, the Gold Bond and Cardinal hide in a small ravine.

As the highway winds northward from Anaconda toward Cripple Creek, the mines of Gold Hill come into view. These mines are also visible from the town of Cripple Creek and from the road adjacent to the Mollie Kathleen. They include the Index, Volcano, Yellow Bird, Mariposa, Conundrum, Midget, Moon-Anchor, Geneva, Jefferson, Anchoria-Leland, Lexington, International, and Grace Greenwood.

A few mines are found outside the normal productive region in the Cripple Creek District. They include the Lincoln, Fluorine, and Galena mines. The Lincoln is north of the accepted productive area while the Fluorine and Galena lie to the northwest.

POVERTY GULCH

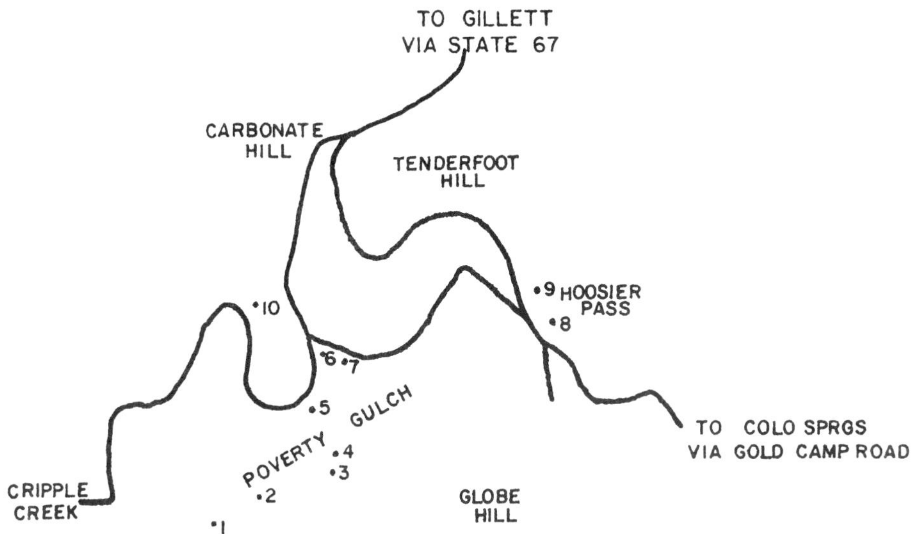

1. ABE LINCOLN
2. CHICAGO AND CRIPPLE CREEK
3. PROPER
4. C.O.D.
5. GOLD KING
6. MOLLIE KATHLEEN
7. BLACK DIAMOND
8. HOOSIER
9. FRIDAY
10. SANGRE DE CRISTO

POVERTY GULCH
NOT TO SCALE

GOLD KING: The Gold King Mine in Early Poverty Gulch (Local History Collection, Pikes Peak Library)

ABE LINCOLN: The W. S. Stratton owned Abe Lincoln first shipped in 1895 when it belonged to the Marinette Gold Mining Company and The Arcadia Consolidated Mining Company. The mine was a good producer, being worked by the Stratton Cripple Creek Mining and Development Company through 1938.

PROPER: The Proper worked in 1904 and in the 1940's with the last apparent production achieved by the Globe Hill Mining Company in 1952 and 1953. It is a Stratton Estate property.

C.O.D.: Charlie Tutt, Colorado Springs real estate man, located the C.O.D. on December 30, 1891. He was joined in the Poverty Gulch adventure in December of 1892 by Spencer Penrose, only six years out of Harvard.

The C.O.D. was owned by the Rebecca Gold Mining Company, Limited, in 1900, after a French firm bought the property for $260,000 but was bankrupted trying to deepen the wet mine. The famous old mine is best remembered as the springboard for the Tutt and Penrose legend and fortunes.

GOLD KING: The direct descendant of the discovery lode of the Cripple Creek Mining District, the Gold King was formerly the Dr. John P. Grannis and Bob Womack, El Paso lode located on the 20th of October, 1890. Shipping first in 1891, the mine produced through 1953 with the last activity occurring in 1958.

MOLLIE KATHLEEN: Located in September, 1891, by Mrs. Mollie Kathleen Gortner, gas and water limited most production from this mine to the period prior to 1900. Shipments continued through 1948 with Mrs. Verner Z. Reed funding extensive work in 1933. Tours started in 1949 and continued through 1960. Ore was being extracted from the 10th and bottom levels during the summer evenings after tourists had left the mine and during the winter. Tours are being conducted at this writing. (1984).

HOOSIER: A prominent shaft locates the Hoosier at the very crest of Hoosier Pass. Production from the twisting Hoosier or Discovery vein continued from 1896 through 1940 when it became the first mine in the area to conduct public tours.

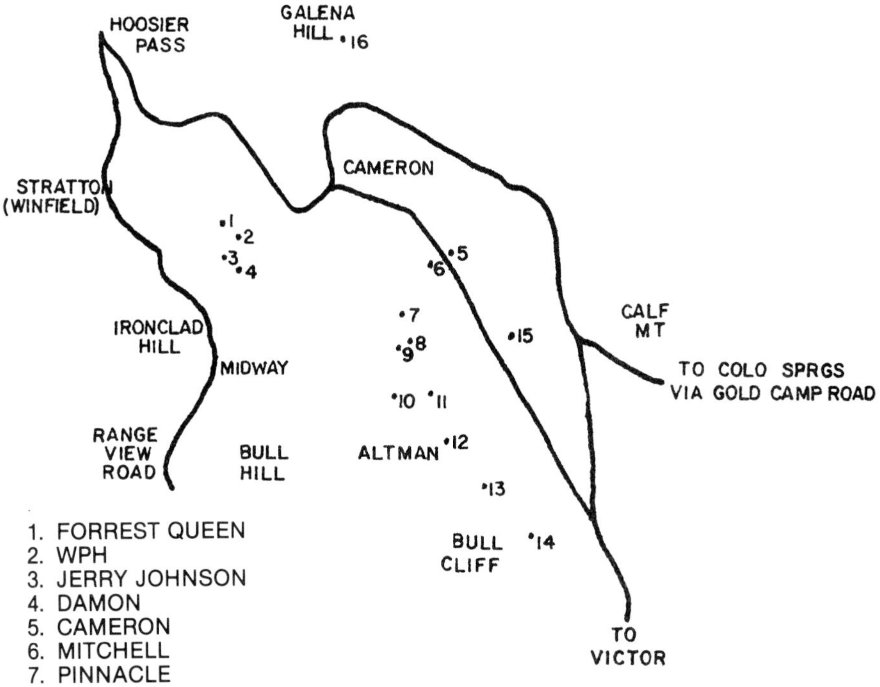

HOOSIER PASS
GALENA HILL •16
CAMERON
STRATTON (WINFIELD)
•1 •2 •3 •4
•5 •6
•7 •8 •9
•10 •11
IRONCLAD HILL
MIDWAY
•15
CALF MT
TO COLO SPRGS VIA GOLD CAMP ROAD
RANGE VIEW ROAD
BULL HILL
ALTMAN •12
•13
BULL CLIFF •14
TO VICTOR

1. FORREST QUEEN
2. WPH
3. JERRY JOHNSON
4. DAMON
5. CAMERON
6. MITCHELL
7. PINNACLE
8. BLANCHE
9. MORNING STAR
10. EMPIRE LEE
11. BUENA VISTA
12. ISABELLA
13. SMUGGLER
14. VICTOR
15. SCHOOL SECTION
16. SUNSHINE

NORTH SLOPE OF BULL HILL AND BULL CLIFF
NOT TO SCALE

Underground workings in the Cripple Creek District (Cripple Creek District Musuem)

NORTH SLOPE OF BULL HILL AND BULL CLIFF

FOREST QUEEN: The mine received some notoriety in 1899, when the wife of Lou Blonger, a prominent Denver underworld figure, forced her husband to default on a $2,000 mortgage she held on the property. Production continued from 1905 through 1912 and from 1917 to 1927 with the Queen Metals Company working the mine from 1927 through 1930, when the Oston Leasing Co. took over. Ore was shipped in 1931 and from 1930 to 1939, when the Lark Mining Co. leased the property. Shipments were made from 1942 to 1944, with the last production recorded in 1952.

WPH: A fabulously rich mine, it produced native gold and oxidized calaverite worth several thousand dollars to the ton. The Woods Investment Company property began shipping in 1903, followed by the the Carlton-controlled United Gold Mines Company in 1928, and finally by Jerry Johnson Gold, Inc. which leased the property in 1935. Production ended in 1941, after an unsuccessful diamond drilling exploration project.

JERRY JOHNSON: By the turn of the century, the Jerry Johnson Mining Company was producing ore from flat clay seams having the appearance and consistency of rich butter. Production continued through 1942, with the last ore shipments occurring in 1948 and 1952.

CAMERON: Controlled by the Woods Investment Company in 1901, Cameron Gold Mines, Inc., produced from the property from 1935 through June 1942, when operations ceased. Golden Cycle acquired the property in 1944, and although intermittently operated through the late 1940's, it became a principal shipper for Golden Cycle.

MITCHELL: It was one of several claims comprising twenty-one acres of Pinnacle holdings. The Pinnacle, itself, was opened by the two Whipp brothers and the two Glenn brothers in the dead of winter, and had produced almost $250,000 from an incline and vertical shaft by 1900. The Pinnacle produced ore worth $342,942.43 in 1904, 1909, 1910, and 1922. Shipments continued from the Pinnacle group from 1928 to 1940, ore being processed in a nearby flotation mill.

EMPIRE LEE: Originally known as the Empire State, the name of this mine was changed in 1923, due to the extensive underground connections with the Lee shaft of the Isabella group. Empire Lee ore sometimes consisted of dump material and stope fillings from 1924 to 1930, when the Carlton-controlled United Gold Mines Company obtained a lease on the property. Production by various lessees continued through 1939. Production continued from the Empire Lee through the better part of the 1940's, with production apparently ending in 1956.

ISABELLA: This mine was the catalyst for the 1894 labor unrest when the mine's over-enthusiastic superintendent, H. E. Locke, served notice that the mine would switch from an eight hour to a ten hour shift but pay the same $3.00 daily wage. The Isabella was then the largest company in Cripple Creek with exclusive ownership of its own claims. The Florence and Cripple Creek Railroad reached the Isabella from the Victor Mine late in 1898, and in December of that year, extremely rich ore was encountered on the 5th and 7th levels of the Lee shaft. On December 29, 1899, the Taylor and Brunton Sampler bought one carload of ore for $219,090.92. This particular lode was found only a few hundred feet below the surface near Bull Cliff.

By 1904, the Isabella shaft was 1,127 feet deep but the productive lode had been lost and the mine was being worked by lessees. The mine had produced over $3,000,000 with the most productive period being from 1895 to 1900. A cyanide mill was erected by the Isabella company in 1906 which treated both mined ore and dump rock until 1918. The ore showed free gold, calaverite, and rusty gold in clay seams which appeared as a brown mud three to four inches thick. Isabella properties were worked continuously until 1923, with the lower levels, initially thought to be barren, producing considerable ore. Ore was shipped in 1933, with United Gold Mines leasing the entire Isabella holdings in 1937.

VICTOR: Discovered in 1891, the Victor Gold Mining Company property was shipping in 1892. David H. Moffat took a bond and lease on the mine and with his business associate, Eben Smith, obtained the claim in 1893 for $65,000. Rich production from the mine began to slacken in 1898. Idle from 1899 to 1903, production began again in 1904, and in 1909 the Roscoe Leasing Company shipped ore followed by almost continuous production to 1924 and some production as late as 1940.

SCHOOL SECTION: This mine was located on block 8 of school section 16 and was known as the Block Eight Mine in the early days of the District. The School Section Leasing Company profitably operated the mine in 1907 and work continued almost uninterrupted through 1926 with almost $1,000,000 being produced by 1922. The Cresson company leased the mine in 1928, and a loading switch was built by the Midland Terminal Railroad for the direct loading of ore. The Kyner Leasing Company shipped ore from 1929 to 1937, subleasing to the Reba-Lee Mining Company for several years. Production continued to 1940 and then proceeded again in 1945 and 1947.

MIDWAY

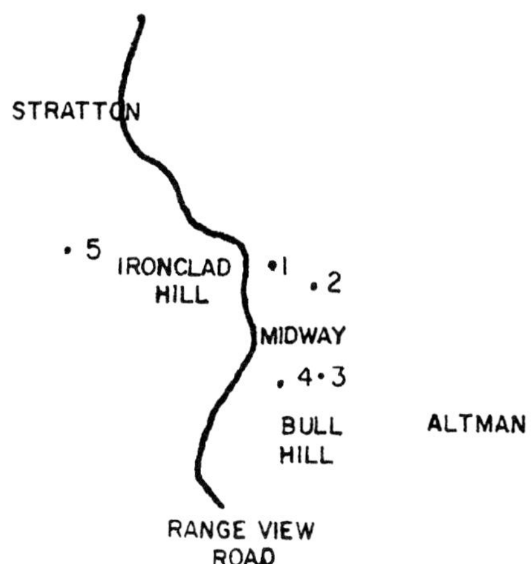

1. SOLOMON
2. PATTI ROSA
3. WILD HORSE
4. GLEASON
5. IRONCLAD

MIDWAY
NOT TO SCALE

SOLOMON: A latecomer to the District, a surface discovery in 1922, it was part of the Damon group belonging to the United Gold Mine Company. It was named after the lessee of the property, Solomon Cain. The mine produced almost continuously until 1934 when the shaft was used by lessees to work the nearby Axtel vein system.

WILDHORSE: Legend says the Wild Horse was discovered and named when a prospector's packhorse dragged him through a rich surface vein in 1897. Regardless of how the ore was discovered, it was bought by the Woods brothers for $35,000 and actively mined until 1903 when gas closed the lower levels. Owned in 1900 by the Consolidated Gold Mines Company, a Woods partnership, the Gleason shaft was located on the 2½ acre Wild Horse claim near the Wild Horse Mine. Although plagued with gas in the lower levels, the Gleason and Wild Horse shafts had produced over $500,000 by 1906. Initially the ore consisted of a fully oxidized clayey mass valued at $300 per ton in some parts of the lode. One observer described the ore as having the consistency and color of rich butter.

On December 27, 1901, Martin Gleason, Wild Horse manager, was thrown down the shaft to his death supposedly after offending a working officer of the Western Federation of Miners. The union, quite naturally, asserted that other mine owners were responsible because Gleason had in his possession valuable papers containing information concerning neighboring property which was in litigation. By 1906, the R E and A Mining, Milling, and Leasing Company was treating ore from the levels which could be mined without the problems of gas. Production continued through 1909, by which time the mine was included in the United Gold Mines holdings.

The R E and A or Wild Horse was dismantled in 1918, and ore values continued to decline. Production had reached over $1,700,000 by 1922, but operators were being forced to install expensive equipment to work the gassy lower levels. Production continued through 1928, the Wild Horse now belonging to the Carlton-controlled interests. Ore was shipped in 1935, 1937, and 1938. The Wild Horse shaft produced a well-nigh nonmillable ore in 1941, with the low grade being so sticky and muddy that the mill closed after trying to process it. The property was inactive in 1942, with the last work being done in 1944.

IRONCLAD: Originally an early day subsurface mine, the Ironclad eventually became a huge glory hole operation as the mine milled more and more low-grade surface ore. The mine was worked almost continuously from 1892 to 1922 and from 1934 until 1941. Operations and ore processors included the Globe Reduction Co., the Cripple Creek Homestake Mining and Reduction Co., and the Kavanaugh Mill. The property is currently leased by the Silver State Mining Company which is leaching gold from the surface ore.

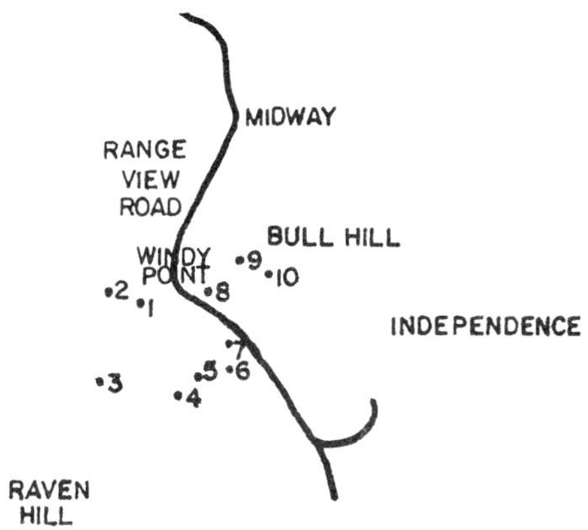

1. AMANDA
2. HAPPY YEAR
3. BLUE FLAG
4. RATTLER
5. SHERIFF
6. BOGART-TWIN SISTERS
7. NIGHTINGALE
8. WAR EAGLE
9. RAMONA
10. FAVORITE

WINDY POINT
NOT TO SCALE

HAPPY YEAR: Early production from the Happy Year included low-grade ore from a 500 foot shaft. Shipments continued in 1909, 1939, and 1940.

BLUE FLAG: This Raven Hill mine was operating as early as 1894, and the Blue Flag Mill, located near the mine itself, was treating the mine's oxidized ore which was averaging $8 per ton in 1908. The Blue Flag Mill was still treating Blue Flag ore in 1909 and 1910, but both the mine and the mill were idle in 1911. The mill was active in 1912 but idle from 1914 to 1918. The Blue Flag-Bogart was working through the Twin Sisters shaft in 1939.

SHERIFF: Located on March 22, 1891, by Matt France and Len Jackson, the Sheriff property had been pockmarked by 1900, with no substantial ore deposits being found. Later, several small deposits of ore worth $500 per ton were shipped by the Sheriff Gold Mining Company. Production continued to be erratic, with ore being shipped off and on between 1904 and 1938.

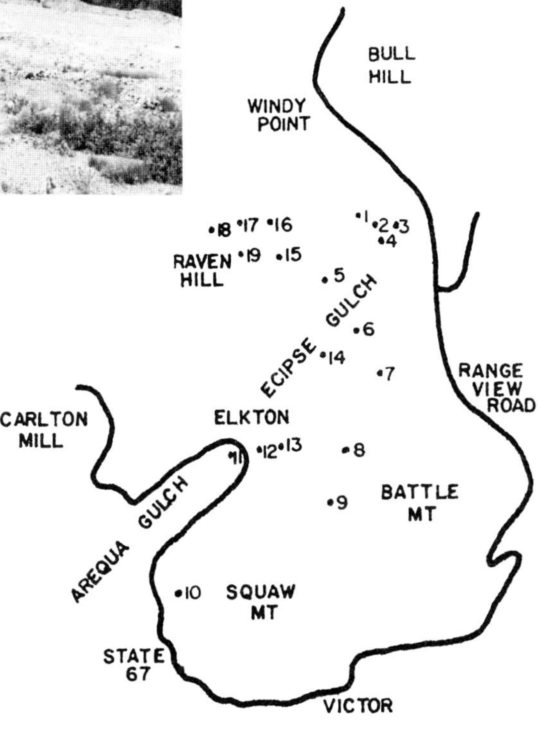

1. MAGGIE
2. GOLD SOVEREIGN
3. DANTE
4. TRILBY
5. CRESSON
6. TRAIL
7. ROSE NICOL
8. OCEAN WAVE
9. CORIOLANUS
10. MAY B
11. SUNSET
12. ECLIPSE
13. CARBONATE QUEEN
14. ADA BELLE
15. MOOSE
16. BERTHA B.
17. JOE DANDY
18. IDA MAY
19. FRANK R

ECLIPSE GULCH
NOT TO SCALE

GOLD SOVEREIGN: This great mine kicked off a profitable career with a 1,000 ton shipment of ore in September, 1899. Production stood at $195,000 by 1900 with the Cripple Creek and Gold Temple Company working the Fox vein through the Jackson shaft in 1904. The Lovett and Whisper veins were also producing, bringing production to $300,000 by 1904. Shipments continued from 1905 onward, with the Union Leasing Company producing ore and doing development work in 1913.

On January 16, 1913, with a great cloud of dust and a rumbling resembling an earthquake, a glory hole over 100 feet deep and 100 feet long appeared as the stope on the 200 foot level of the mine dropped overlying ground into the hole. The cave-in of the worked-out slope, which had reached within 100 feet of the surface, swallowed equipment and some surface structures but no one was hurt.

Ore was shipped from 1914 to 1924 and in 1925. The Gold Sovereign dump was being worked in 1926 with work continuing through 1931. Lessees operated through the Gold Sovereign shaft in 1935 with lessees making good production to 1938. By 1939, the Cresson Consolidated Gold Mining and Milling Company owned the property and was actively working it. Work continued through 1941, but the mine was shut down in 1942 due to a shortage of lessees.

DANTE: It was the largest true "pipe" in the District, the ore deposit overlapping the so-called Cresson "blowout." Production started in 1899 and continued through 1910 with total production standing at $748,736. The Dante produced almost continuously until 1930, coming under control of the Cresson company in 1927. Several mills, including the Dante or Gaylord, Reed, Big Toad, and Dante, processed ore from the workings. The last production was from 1935 to 1939.

TRILBY: The small 4/10 acre Trilby claim was one of the first shippers in the District. The property contributed measurably, along with the Moose Mine, to the turn of the century production for the Moose Gold Mining Company of $145,000. Shipments continued from 1907 through 1913 with the Trilby Mill working in 1908 and 1909. The property became part of the Cresson in 1914 and continued to produce for that company.

CRESSON: When the Cresson was bought by J. R. and Eugene Harbeck in 1894, the Chicago insurance men were sure they had a lemon on their hands. Around 1910, Richard Roelofs, a heretofore unimpressive civil engineer who had a penchant for making marginal mines pay, began deepening the mine, mining huge amounts of low-grade ore from even larger stopes, and erecting an aerial tram down Eclipse Gulch. Roelofs, with the help of the Roosevelt Drainage Tunnel, soon dispensed with the mine's debt, production reaching tens of thousands of dollars per year. The real payoff came on November 24, 1914, when miners on the 12th level broke into a cavern or vug, 40 feet high, 20 feet long, and 15 feet wide. The walls were covered with sylvanite and calaverite crystals with flakes of pure oxidized gold as large as thumbnails. Vault-like doors were installed, guards were posted, and miners personally picked by Roelofs himself merely scraped the rich ore from the walls and sacked it. The vug produced $1,200,000 in four weeks with the production for 1915 alone reaching $2,000,000.

A. E. Carlton bought the mine in 1916 for almost $4,000,000 and the Cresson soon became the most important producer in the District. Production slumped in 1920, and the lower levels below the drainage level of the Roosevelt Tunnel were abandoned in 1929. The Carlton Tunnel unwatered lower levels in the mine in 1941, but lack of materials and manpower shut down the mine in 1943. The Cresson machine shops were converted to manufacturing some small essential war items. Work started again in April, 1946, but production was interrupted when a huge hole opened near the shaft, swallowing the blacksmith shop, warehouse, and upper end of the tramway.

The Cresson, which eventually came to develop properties such as the Dante, Rose Nicol, Gold Sovereign, and Portland, slumped again in 1949. Production resumed on a much lower level in 1951 and continued on a much subdued level until 1961, when the mine closed after reaching a total production of $51,000,000. The Cresson was acquired by the Golden Cycle Corporation in 1971 through foreclosure proceedings. By the summer of 1981, a proposed 500 foot incline west of the Cresson shaft had been driven 210 feet toward an estimated 150,000 ton ore deposit.

ROSE NICOL: Production here started in 1909 with the Rose Nicol Gold Mining Company working the property almost continuously until the United Gold Mines Company took over in 1929. Production continued through the Trail Tunnel, the Cresson, and the Rose Nicol itself until 1948 with the last production coming in 1953.

MAY B: It was an early producer for the Columbine Victor Deep Mining and Milling Company, a Woods-controlled company. Production continued in 1910 and 1913 with the last work done in 1935.

ECLIPSE: A pioneer producer, the Eclipse shipped ore consisting of rusty oxidized free gold in sand quartz and fluorite from 1892 to 1894. Plagued by gas and water problems, the workings had produced about $50,000 by 1900. Because of the close proximity of the Eclipse and Carbonate Queen shafts, which were connected, this mine was also called the Queen.

ADA BELLE: Located near the Cresson, the Ada Belle produced in 1935 and 1937.

JOE DANDY: Born with the creation of rich gold deposits in the Cripple Creek Mining District, the Joe Dandy shipped good ore in 1901 and produced a fortune for James H. Peabody, who as governor of Colorado in 1903, sent the Colorado Militia to riot-torn Cripple Creek. The mine was idle in 1903 and 1904, but in 1906 a three inch vein with ore averaging $1,000 per ton was found on the first level, only 180 feet below the surface. To process the ore, the company bought the cyanide mill of the Cripple Creek Cyanide Company at Gillett and moved it to the mine site. Ore shipments continued almost uninterrupted through 1911 with the Joe Dandy dump giving up low-grade ore. Production by the Joe Dandy company and lessees was only marginal when the mill was dismantled in 1916. Much of the work in the 1920's was done by the Red Raven Gold enterprise.

The Joe Dandy Company was officially incorporated in 1925 with the mine being reopened in 1928 and shipping in 1929. The New Zealand Gold Mining Company opened up a good body of ore here in 1934 with shipments continuing in 1935. The Joe Dandy property produced continually from 1936 until 1947 with the Joe Dandy Mining Company doing the last work in 1950.

Other mines in Eclipse Gulch include the Maggie, a Cresson property; the Trail, last worked through the Cresson in 1936; the Ocean Wave, a Creede and Cripple Creek Mining and Milling Company property; the Coriolanus, which was worked through a crosscut from the Ajax in 1937; the Sunset, which merged with the Eclipse; the Carbonate Queen, located above the Eclipse and last worked in 1945; the Moose, a rich but gassy mine; the Bertha B., which encountered extremely rich ore by 1945; the Ida May, an Elkton and then later an United Gold Mines property; and the Frank R.

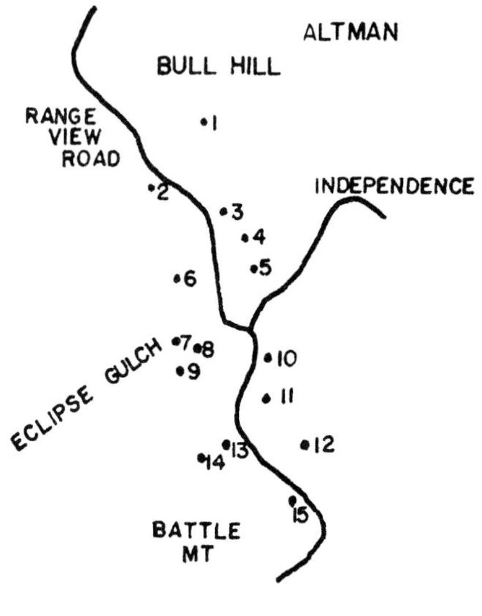

VICTOR

1. AMERICAN EAGLES
2. LOGAN
3. ORPHA MAY
4. RUBIE
5. SACRAMENTO
6. BLUE BIRD
7. DEXTER
8. LOS ANGELES
9. COLORADO CITY
10. LAST DOLLAR
11. MODOC
12. CLYDE
13. WISCONSIN
14. PORTLAND NO. 3
15. RIGI

SIX POINTS

NOT TO SCALE

AMERICAN EAGLES: This mine, highest in the District, was purchased in 1895 by W. S. Stratton. The American Eagles group, actually three shafts, was connected to the John A. Logan. The American Eagles had reached 1,540 feet by 1902 making it the deepest shaft in the District. Work continued in 1904 and from 1915 to 1922 with the dump shipping in 1923. After a long period of idleness the shaft was reopened in 1924 by the Stratton Leasing Company.

Despite extensive development in 1926, work and shipments continued from the American Eagles only until the end of the year. Ore was shipped in 1935, and in 1936 a rich strike was made between the 19th and 21st levels, producing ore worth $7.00 per pound. Still the highest working mine in the Cripple Creek District, it continued to produce ore from 1937 through 1940.

JOHN A. LOGAN: More commonly called the Logan, this mine was producing in 1893 and was purchased in 1895 by interests of W. S. Stratton. The mine was so high it was water free but the mine was closed in 1904 due to gas so dense that it appeared as a bluish haze. Two men were overcome and killed by this gas in 1901 and another succumbed in 1915.

Production in the mine was marginal by 1905 with the mine working in 1919 and 1920 and the dump shipping in 1930. International Gold Producers, Inc. worked the Logan in 1933 and 1934. The Stratton Estate-owned property was worked in 1935 and was producing one hundred tons of ore daily in 1936. Lessees continued to produce from the dump and shaft in 1944 with work occurring as late as 1947.

ORPHA MAY: Eventually 1,264 feet deep, the Orpha May, once owned by the Union Gold Mining Company, later became the property of the Stratton Estate. The mine was closed in 1904 and only marginally productive by 1905. It was a working Stratton Estate property from 1915 through 1919 and again in 1921 with the dump producing ore from 1922 through 1925. The Stratton Lease worked the Orpha May in 1927 and 1928. Ore was shipped in 1930 with lessees working the Mountain Beauty property through the Orpha May shaft in 1935. Work continued as late as 1947.

RUBIE: In 1900, the Princess Alice Company had a lease on the four-acre Rubie claim which was owned by the Rubicon Gold Mining Company. The mine continued to ship from 1908 through 1913 with the Rubie or Worcester Mill working in 1916 and 1917. The dump gave up ore in 1927 and 1928 with production shown for 1935, 1936, and 1952. It was worked intermittently by a local lessee in 1974 and by 1981, Gold Ore Limited of Cripple Creek, the new owner of the property, was operating a leaching pad at the site.

SACRAMENTO: The Sacramento Gold Mining Company produced a net value of $7,000 from ore mined prior to 1899 with $10,000 worth of gold given up by 1901. Little production was achieved after this date and the mine became part of the Stratton Cripple Creek Mining and Development Company. There was some production in 1938.

BLUE BIRD: The highly mineralized Blue Bird shipped ore consisting of oxidized free gold, tellurides, massive calaverite, and an unusual amount of silver running several hundred ounces per ton. Most of the production was accomplished in 1893 and 1894 with $300,000 accumulated by 1900.

The mine had reached its prime by 1906 with gas closing the mine at times. One man was killed by gas in 1905 and three in 1906 as the Blue Bird Mining and Milling Company property reached a depth of almost 1,400 feet. The mine was worked almost continuously until 1941. Explosives were stored in the explosives igloos which are still visible at the site.

LOS ANGELES: An early producer, the Los Angeles had recorded $100,000 worth of shipments prior to the turn of the century. Little work was done after 1901, the Los Angeles Gold Mines Company eventually becoming a Stratton Cripple Creek Mining and Development Company property. Ore was shipped in 1905, 1927 through 1930, 1935 through 1939 and in 1944.

LAST DOLLAR: The Last Dollar Gold Mining Company worked this property as early as 1896 and had produced $700,000 by 1900. Production continued through 1906 with gas becoming a problem on the 10th level, miners developing terrific pains after a few days work. The Last Dollar and Modoc veins produced almost continuously from 1908 until 1925 when the mine was purchased by the Portland Gold Mining Company. A connection was made between the Portland and Last Dollar, and the Wisconsin vein, long a bone of contention between the two mines, was exploited.

The Los Angeles Mine was worked through the Last Dollar in 1928 and the dump was worked in 1930. In 1933, The United Gold Mines Company acquired the mine through the purchase of the Portland and ore was shipped until 1940.

MODOC: This old property had produced from the Ocean View claim to the tune of $65,034.91 by 1899 and $500,000 by 1904. Ore was shipped from 1908 until 1928 when a large vein gave up sylvanite worth $140 to $240 per ton. By 1935 the Modoc was part of the Portland system and a United Gold Mines property.

CLYDE: A marginal mine, the Clyde promised potential riches but never quite produced the expected bonanza. The barn red surface buildings are clearly visible from the Range View Road on the north side of Battle Mountain.

PORTLAND #3: Lessees worked through the Portland No. 3 shaft starting in 1935. It was not as famous as its sister shafts, the Portland No. 1 and No. 2, located over the crest of Battle Mountain.

RIGI: Working as early as 1897, the Rigi group was leased by the North American Gold Mining and Development Company from the Rigi Group Gold Mining Company, Limited, of which the Earl of Essex was president and chairman. The three shafts on the property produced in 1900, 1909, 1911, and 1936.

Several other rich properties were located in the Six Points area including the Dexter, the Colorado City, which became a ventilation shaft for the Portland Mine; and the Wisconsin, finally mined by the Portland in 1925.

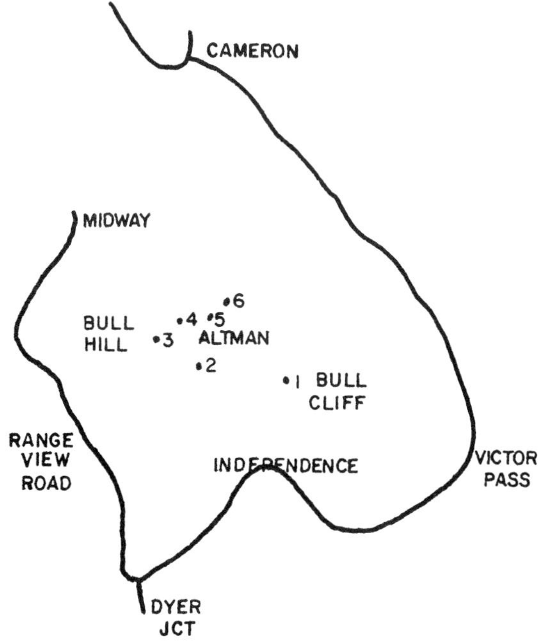

1. DEADWOOD
2. SOUTH BURNS
3. ZENOBIA
4. PHARMACIST
5. BURNS
6. PINTO

ALTMAN
NOT TO SCALE

DEADWOOD: The Woods-controlled New Zealand Consolidated Mining Company consisted of the New Zealand, Pauper, and Deadwood claims. Oxidized ore and claylike seams of rusty gold produced fabulous returns, often being picked out by hand. By 1910, the claims were part of the United Gold Mines firm.

From 1921 until 1931, lessees worked several claims of the Deadwood. Gold Bullion Mines, Inc. worked the North or Upper Deadwood in 1935 while Blackwood and Associates worked the South or Lower Deadwood. Shipments occurred almost continuously through 1941 with some activity as late as 1958.

SOUTH BURNS: Formerly the property of the Calumet Mining and Milling Company, the mine became the property of the Acacia Gold Mining Company in 1895 and was sometimes referred to as the Acacia. The South Burns was worked by the Acacia company as late as 1926.

The Nuestra Ventura Mining Corporation worked the mine in 1927 and 1928. The same properties continued to produce for the Acacia Gold Mining Company through 1936. Golden Conqueror Mines, Inc. opened a large body of ore and was shipping from the South Burns shaft in 1937 with the Acacia producing from 1938 to 1940. The South Burns shaft was used to reach deposits in the Shurtloff and Findley mines in 1941 with the Acacia producing almost continuously until 1947.

INDEPENDENCE

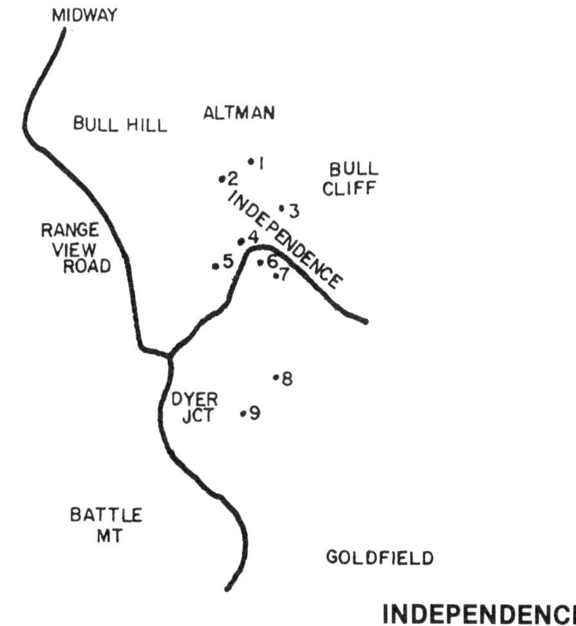

1. TRACHYTE
2. SHURTLOFF
3. DELMONICO
4. ATLANTA
5. FINDLEY
6. HULL CITY
7. GLORIETTA
8. LONGFELLOW
9. SITTING BULL

INDEPENDENCE
NOT TO SCALE

SHURTLOFF: Miners from the Shurtloff, which was owned by the Findley Gold Mining Company, were saved from being involved in the Independence station demolition in 1904 when their work shift caused them to arrive at the station late. The mine produced intermittently from 1905 to 1916, becoming a United Gold Mines property in 1930. Production continued, sometimes with the Golden Conqueror lessees working through the South Burns shaft, until 1942 when the war drained labor from the area.

DELMONICO: This mine was owned by the Union Gold Mining Company in 1900 and, along with the Pikes Peak and Orpha May mines, had earned $1,000,000 for the company. A Stratton Estate property in 1904, ore was running one to two ounces of gold per ton. The mine was grouped with the Free Coinage Gold Mining Company in 1923. Production occurred from 1935 until 1941.

ATLANTA: The Atlanta before the hoist house was razed.

HULL CITY: The Independence Town and Mining Company and lessees had produced almost $1,000,000 from the Hull City by 1900. Water became a real problem by 1904, and the lower two levels were allowed to flood. The Vindicator controlled this mine by 1906 with work continuing to produce a lowering grade of ore as deeper levels were being opened by 1909. Shipments continued in 1911, 1916, and 1922. Hidalgo Gold Mines, Inc. leased the United Gold Mines property in 1934 and 1935 with production continuing until 1941 when a new ore house was constructed. Work stopped in 1942 due to the manpower shortage, but some production was recorded when the dump shipped in 1953 and 1958.

GLORIETTA: By 1904, the Independence Consolidated Mining Company had realized close to $2,000,000 from the Hull City. The King, Vaughn or Glorietta shaft produced calaverite, some oxidized ore, and a large amount of silver.

LONGFELLOW: This mine was on the scene in 1895 with W. S. Stratton acting as president, owner, and operator of the Longfellow and Longfellow No. 2 claims, about eighteen acres in 1900. The Longfellow shipped in 1919 and 1920 and from 1936 to 1938. Production continued in 1940, and the mine was worked in 1941 by C. E. Sullivan and the Longfellow Mining Company. The last production was in 1942.

SITTING BULL: Located on the Sitting Bull claim, this mine was owned by the Keystone Mining and Milling Company which had produced about $10,000 by 1899.

The community of Independence is also the site of the Trachyte, owned by the Woods company and then United Gold Mines and the Findley, a huge two-shaft mine which lost thirteen men in the Independence depot bombing.

GOLDFIELD

BULL HILL
ALTMAN
BULL CLIFF
INDEPENDENCE
VICTOR PASS
RANGE VIEW ROAD
BATTLE MT
GOLDFIELD
268
BIG BULL MT
VICTOR
HOLLYWOOD

1. VINDICATOR NO. 2
2. VINDICATOR
3. LILLIE
4. ANNA J
5. GOLDEN CYCLE
6. THERESA
7. GOLD KNOB

GOLDFIELD
NOT TO SCALE

GOLD KNOB: The Gold Knob Mining and Townsite Company owned most of the townsite of Goldfield. Ore consisted primarily of oxidized free gold with a little calaverite and native free gold. The Gold Knob was acquired by the United Gold Mines Company in 1937.

VINDICATOR: By 1900, Vindicator production stood at $2,000,000. Water was initially encountered at 500 feet, and in 1903 the pumping operations in the deeper Vindicator were evidently unwatering the Findley, Golden Cycle, and Hull City mines, discharging 300 to 500 gallons of water per minute. The water was used by the town of Victor, being transferred directly from the Vindicator pumps.

On November 21, 1903, Harry Orchard, former high-grader turned terrorist, inadvertently killed Vindicator superintendent Charles H. McCormick and shift boss Malverne Beck when his dynamite charge exploded on the sixth level of the mine. He had intended to kill fifteen scab workers when they stepped from the cage on the seventh level but had confused the level on which he placed the bomb.

Production for the preceding twenty years stood at millions of dollars by 1914. The Vindicator took over the Golden Cycle in 1915, and in 1916 the Vindicator was looking at her dump reserves as mine ore was becoming low grade. By 1920, ore reserves were diminishing and when pumping was temporarily discontinued, water rose quickly up the shaft. In 1921, a tunnel had to be driven 3,000 feet to the Roosevelt Drainage Tunnel to unwater the mine. On June 9, 1922, the Vindicator and Golden Cycle were bought by A. E. Carlton's United Gold Mines Company.

Production continued through the war years and by 1951 over $27,000,000 worth of ore had been mined. The dump or mine was producing until 1959, when the United Gold Mines property was a major producer. The last producing level was at about 2,200 feet with only limited production, most of the ore coming from eight to ten higher veins including the La Bella and Middle.

GOLDEN CYCLE: The Golden Cycle Mining Company was organized in November of 1895 to take over direction of the mine. It had produced over $1,500,000 by 1900 when John T. Milliken, the forty-seven year old St. Louis chemical magnate, bought the Golden Cycle, Anaconda, and other mines from David Moffat for $1,700,000. In March, 1915, Milliken sold 95% of the capital stock to A. E. Carlton, putting the Golden Cycle under control of the Vindicator for around $1,500,000. The Anna J shaft of the Golden Cycle produced as late as 1940.

THERESA: The Theresa Gold Mining Company worked the mine almost continuously from 1895 to 1900 when the property was opened to lessees. The mine produced in 1904 and 1905 with the United Gold Mines Company working the property from 1923 to 1933. The impressive steel headframe and bins were constructed in 1934 and 1935 thus eliminating the constant fire hazard and most of the surface jobs as well. The mine continued to produce from 1941, when it was considered part of the Vindicator group, through the war years, shipping a large amount of Golden Cycle-Theresa dump ore.

The United Gold Mines Company was rehabilitating the Theresa Mine to reach new areas for mining and prospecting in 1960 and shipments were made in 1961.

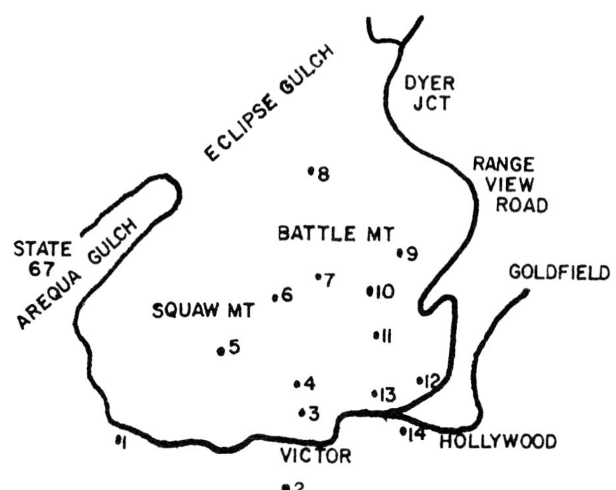

1. SANTA RITA
2. ST PATRICK
3. GOLD COIN
4. MARY CASHEN
5. NELLIE V.
6. AJAX
7. GRANITE
8. LOST ANNA
9. PORTLAND NO. 2
10. PORTLAND NO. 1
11. INDEPENDENCE NO. 2
12. INDEPENDENCE NO. 1
13. STRONG
14. JOLLY TAR

BATTLE AND SQUAW MTS
NOT TO SCALE

On the 1000′ level of the Portland#1.

GOLD COIN: Frank Woods, of the Woods Investment Company, noticed a twenty-inch wide vein while excavating the basement of the Victor Hotel in March, 1894. He traced the vein to the Gold Coin claim and took a $50,000 lease on the property. When fire destroyed the shaft house in 1899, it was rebuilt using ornate pressed brick and stained glass windows. Dump ore was hauled through Squaw Mountain via the Columbine Tunnel to the Economic Mill to prevent encroachment of waste rock on the city of Victor.

Charles MacNeill, Spencer Penrose, and Charles Tutt bought the Gold Coin in 1903, the mine having produced $6,000,000 by 1904. When a lower level of the mine was unwatered, a thick mud two feet deep was found to contain two ounces of gold per ton. The mine worked from 1913 through 1917 and was 1,300 feet deep in 1926. The mine was worked in 1935, and the Golden Cycle Corporation bought the Gold Coin along with the Granite, Ajax, and Sunshine Sedan from the International Gold Mining Company for $80,000.

NELLIE V.: The Nellie V. Gold Mining Company, consisting of the War Eagle and Nellie V. claims, which were unwatered by the Columbine Tunnel, had produced $80,000 by 1900 from rich ore. The Squaw Mountain Gold Mining Company worked the properties in 1925 and the Granite Gold Mining Company leased the property in 1926 and 1927. The Nellie V. and War Eagle properties worked in 1941.

AJAX: Owned by the Ajax Gold Mining Company, this famous mine shipped granite honeycombed with gold tellurides as early as 1895. About 1903, Charles Tutt, Spencer Penrose, Charles MacNeill, Clarence C. Hamlin, and Oliver Shoup owned the mine, with shipments continuing through 1912 when the Colburn-Ajax (Portland) Mill was working on the site. Ore shipments continued until August, 1922, when the mine was purchased by the Ajax-Tonopah Mining Company. The mine was worked until 1925 when it came under control of the Granite Gold Mining Company. The New Market vein was particularly productive, the mine being reopened through the Dillon shaft.

The Ajax came under control of the International Mining Corporation in 1933, and a crosscut was completed from the Portland No. 2 shaft at the Roosevelt Tunnel level. Golden Cycle purchased the mine in 1935.

Production continued but unwatering the lower levels was causing the mine to operate at a loss. Operations were curtailed in 1940 while the property awaited the arrival of the Carlton Tunnel. The lower levels were allowed to flood and work was limited to surface projects and mining above the 2,000 foot level. As the Carlton Tunnel unwatered the lower levels, a higher grade of ore became available and shipments continued through the war years period. The Ajax and its satellite properties became the seventh largest gold producing mine in the United States in 1947. The mine ceased operations in 1949, rehired crews in 1950, and resumed mining in 1951, becoming the largest producer in the District in 1953, and the largest producer in Colorado the following year. Mining continued, a steel headframe being erected in 1959, but with the Carlton Mill closing, production ceased on December 31, 1961.

The Cripple Creek Gold Corporation, a subsidiary of Golden Cycle Corporation, announced a $6,000,000 rehabilitation program in 1972 and followed it up in 1973 by constructing a new hoist, retimbering the shaft, and cleaning up the Carlton Tunnel. In 1974 the shaft was being deepened with a drilling program showing minable veins in the New Market system, these ore bodies being pursued in 1975. The shaft had been repaired to the 3,100 foot level by 1976, the repairs costing $300 per foot and taking ten months. The Cripple Creek and Victor Gold Mining Company continued to work in the mine repairing shaft guides in 1981 and working the 3,100 foot, 3,200 foot, and 3,350 foot levels, while stockpiling ore until the Carlton Mill was ready to process it.

PORTLAND: James Doyle and James Ferguson Burns located the Portland claim on January 22, 1892, after some claim shifting on Battle Mountain left this small piece of ground vacant. The claim was only 69/1000 of an acre. The mine, which they called the Portland after their home in Maine, came to include 180 acres on Battle Mountain, employed 700 miners on three eight-hour shifts, and had one of the deepest shafts in the District at 3,200 feet. It was one of the greatest producers in the Cripple Creek District.

The Florence and Cripple Creek Railroad reached the mine in 1896, the railroad having to build a 242-foot steel tunnel which was eventually covered by the Portland dump. Production from the Diamond, Captain, and Buckeye veins was totaling over $1,000,000 a year. The mine was the only one not to close because of labor troubles in 1903, but the Colorado Militia forced the mine to shut down in 1904.

Two mills, the Portland in Colorado Springs and the Victor or Portland near the No. 3 shaft on Battle Mountain, processed ore from 1908 to about 1921. Operations were reduced in 1909 while waiting for the Roosevelt Drainage Tunnel to start unwatering the mine. $60,000 spent on the drainage tunnel was rewarded when the water level began to fall in the Portland in 1910. Despite the inflationary period during World War I the Portland was able to buy the Independence Mine in 1915.

The Independence Mill, acquired when the Portland purchased the Independence, was connected to the Portland No. 1 shaft as well as to several dumps in the area. An eight-ton electric motor unit and ten four-ton cars made a round trip from the chutes at these locations to the mill in ten minutes. The Independence Mill was shut down in 1928.

The Roosevelt Drainage Tunnel was connected to the No. 2 shaft in 1918, but a program to train men from agricultural areas to be miners fell through. 1920 was a bad year due to low ore reserves, low manpower, debt and a missed dividend. Things got better in 1921 while the Cresson and Portland repaired the Roosevelt Tunnel with ten-inch reinforced concrete arches and 2,500 feet of track which had to be laid underwater. The Portland bought the Last Dollar in 1925 and drove a connection to the mine on its 17th level. Things were looking good for the Portland.

The Portland started downhill again about 1927. The Last Dollar work ceased and crosscuts into the Strong proved unproductive. To save money, pumps were shut down, flooding the lower levels. By 1934 the Portland had over seventy-four miles of underground workings and had produced over $60,000,000 worth of ore. A break-even operation, the Portland was bought by the United Gold Mines Company and operations continued with the owners supervising and repairing and running the shafts and lessees doing the actual mining. The Ajax and Cresson pumping operations kept the water level down and a new steel ore house was started at the No. 1 shaft. By 1941, the Carlton Tunnel arrived under the Portland No. 2 at the 3,252 foot level. When the 800 feet separating the tunnel from the shaft was excavated, 125,000 gallons of water rushed through the Carlton Tunnel, destroying track and depositing debris in the tunnel.

The Portland continued to work through the war years, with two incomplete laterals, one from the Cresson-Rose Nicol property and another from the Vindicator, approaching the Portland No. 2 in 1949. The mine produced in 1953, and in 1971 the Golden Cycle Corporation acquired the mine through foreclosure proceedings.

INDEPENDENCE: W. S. Stratton, the thin, forty-two year old, white haired Colorado Springs winter carpenter and summer prospector, staked the Independence claim on July 4, 1891, enclosing a granite ledge on the south slope of Battle Mountain. He had more faith in his Professor Lamb and Washington claims, however, optioning the Independence to L. M. Pearlman in 1893. Pearlman overlooked the rich ore that Stratton had found in the mine after taking the option, but Stratton had to sweat out the thirty day inspection period in total apprehension.

Stratton decided to limit his production to $2000 per day and drew ore from the mine like money from a bank. By 1895, the Independence was the most profitable mine in the District having expanded to eleven adjacent claims. By 1899, the mine had shipped almost $500,000 worth of ore with $7,000,000 blocked out for shipment.

In 1899, Stratton sold his wonderful mine to the Venture Corporation of London for $11,000,000. One million dollars went to the salesman, Verner Z. Reed, and Stratton had the sales proceeds plus the estimated $2.5 million clear profit on ore already removed from the mine. Stratton's Independence, Limited, proceeded to gut the mine but lowering ore values with depth, water, and labor problems (fifteen miners were killed on January 26, 1904, when the cage they were using dropped to the bottom of the shaft) combined to lower production.

The mine and mill of the Independence were sold to the Portland Gold Mining Company on June 30, 1915, for $372,000. The mine had produced over $23,000,000 with the Indepence Mill working since 1908. The mine became a United Gold Mines property in 1934 with the purchase of the Portland. The mine was not profitable for the company in 1935 but lessees continued to find ore in the old mine until 1938.

STRONG: Sam Strong, a Colorado Springs lumber hauler and general roustabout, staked the mine in 1891, but had optioned it to William Lennox, Judge Ernest A. Colburn, and Ed Giddings, all of Colorado Springs, for $60,000. These men accused Strong of wrecking their lease when the Strong shaft house and hoist were blown up by striking miners in May, 1894. Strong, who was later cleared, was killed in a barroom shoot-out in 1901.

Production was almost continuous to 1911 when the Strong began to be unwatered by the Roosevelt Tunnel. Ore was shipped from 1912 through 1924, in 1929 and 1930, from 1932 through 1942, in 1946, and by the Front Range company in 1947 and 1948. Front Range Mines, Inc. shipped concentrates to their mill in Leadville in 1950, 1951 and 1953. Production continued in 1956 with 100 ounces of gold produced in 1957. In 1974, the Strong shaft was cleaned out and the mine extensively sampled. Production was probably around $13,000,000.

The Battle and Squaw Mountain area was also the site of the Santa Rita, owned by the Atlantis Mines Corporation; the St. Patrick, in downtown Victor; the Mary Cashen, worked by the Front Range company; the Granite, located between the Ajax and Portland; the Lost Anna, a Portland shaft; the Independence No. 2, a second shaft of the famous Independence; and the Jolly Tar, part of the Ajax.

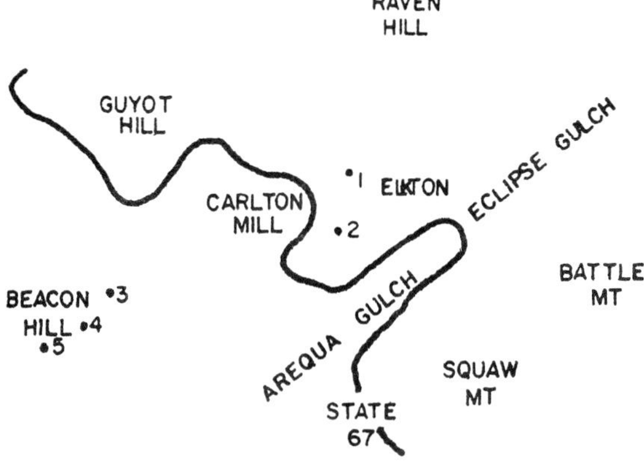

SOUTH SLOPE OF RAVEN HILL AND EAST SLOPE OF BEACON HILL

NOT TO SCALE

1. ELKTON
2. THOMPSON
3. GOLD DOLLAR
4. PRINCE ALBERT
5. BEACON

The Elkton Hoist.

SOUTH SLOPE OF RAVEN HILL / EAST SLOPE OF BEACON HILL

ELKTON: Supposedly named for a pair of bleached antlers found on the claim, the Elkton was staked in 1891 by a Colorado Springs blacksmith, William Shemwell. He sold a half interest to Colorado Springs grocer Sam Bernard for dropping a grocery bill. Bernard and his brother George eventually gained control of the mine, developed the mine for two years and, on the edge of bankruptcy, made a rich strike which brought in $40,000 the first week. The Elkton and Kentucky Bill were joined by the Appie Ellen No. 2 in 1898, the Thompson in 1899, and the Raven, Tornado, Gregory, Catherine and Walter claims in 1900. The Bernards, millionaires by 1902, retired. They died in the 30's, relatively poor.

By 1903, the El Paso Drainage Tunnel was unwatering the Elkton, but gas became a problem on levels below the sixth. The Roosevelt Drainage Tunnel helped unwater the 8th and 9th level in 1910 and reached the property by 1915. Even so, ore deposits were exhausted and the mine was declining by 1917. On May 20, 1923, the Elkton Mining and Milling Company turned operations at the mine over to the First National Bank of Cripple Creek which operated the mine for the remainder of the year. The mine worked from 1924 to 1927 with gas becoming a real problem in March of 1927. Owned by the Carlton interests by 1928, the mine produced on a reduced scale until 1935.

The Raven and Beacon Hill Gold Mining Company was acquired by the Elkton Company in 1936, and shipments continued until 1940. Evidently no production was made during the war years, but the property shipped in 1946, 1948, and 1949 with the last shipments being dump ore in 1956. The picture shows the mine at its peak of productivity.

GOLD DOLLAR: The property was a Woods mine in 1904 and worked in 1907 and 1909, with the Mabel M contributing to production in 1910. By 1911, the Roosevelt Tunnel had almost completely unwatered the mine. The Mabel M was being shipped again by lessees in 1912, and the Gold Dollar continued to produce from 1913 through 1918. The mine worked in 1923 and made a small shipment in 1924. Shipments continued from the mine and dump in 1926, 1927, and 1928 when it became a Carlton interest-controlled mine. Gold Dollar continued to produce from 1929 to 1932, with the New Gold Dollar Mining Company producing from 1933 through 1937, and some production occurring as late as 1945.

PRINCE ALBERT: The mine began shipping in 1893, but by 1894 it was a disorganized mess due to the irregular drifting and stoping done to follow the twisting vein. Originally located by the unpredictable Irishman Tim Hussey, he reportedly went insane when he was relieved of his mine by less than honorable means. By 1900, $500,000 worth of ore had been produced. The Beacon shaft is pictured.

The mine was worked on a small scale from 1904 through 1906 and was now part of the Prince Albert Mining Company, Limited. Shipments of ore from 1908 through 1920 created huge stopes 30 feet wide and 200 feet high. The Prince Albert dump worked in 1925, and the mine produced some ore between 1935 and 1943.

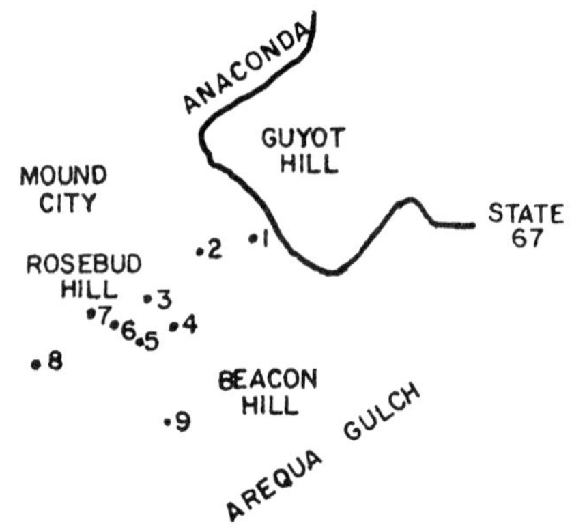

1. NICHOLS
2. MAID OF ORLEANS
3. CK&N
4. LITTLE MAY
5. EL PASO
6. OLD GOLD
7. HENRY ADNEY
8. MARY NEVIN
9. HIAWATHA

BETWEEN BEACON HILL
ROSEBUD HILL
NOT TO SCALE

THE EL PASO.

EL PASO: Sam Bernard of Elkton fame controlled the El Paso in its early days. In 1894, the Jewelry Shop stope on the 2nd level was discovered. It eventually produced $5,000,000 worth of gold, $75,249 of it in one 2,902-pound carload of ore in 1904. It took two hundred men ten years to deplete the stope of its treasures.

On September 6, 1903, the El Paso Drainage Tunnel was connected to the mine and it was unwatered at the rate of fifteen feet a month. On March 26, 1906, an underground torrent of water entered the mine at the 4th level. Men working below that level came close to drowning as the drainage tunnel, which also entered the mine on the 4th level, could not handle the flow. At 5,000 gallons per minute the water soon submerged two-thirds of the mine, rising 600 feet in a little over two hours. El Paso stock lost much of its value, and pumps worth $40,000 were left useless below the water.

By 1915 the company was $70,000 in debt and working on credit. Lessees continued to work the property with the mine closing several times until 1933 when the New El Paso Mines company began production. The mine eventually fell under control of Hidalgo Gold Mines, Inc. This company then merged with the New El Paso Gold Mines Company to form Gold, Inc. Production continued in 1936 and off and on through 1948 when the mine was again closed after $20,000,000 worth of gold had been dug out of thirty-six miles of underground workings. Production resumed in 1950 and again in 1956. A Colorado Springs syndicate then purchased the mine for scrap metal but ended up planting the 2nd level in mushrooms, mining ore on the 9th, and giving tours on the 3rd. The El Paso was probably one of the few mines to be considered by the Defense Department as an underground control site.

The Nichols Mine, part of the El Paso, is shown above.

MARY NEVIN: This little mine produced in 1913 and 1914 and from 1937 to 1941, 8,250 ounces of gold being found before the mine was idled in the early 1940's. The mine was sold to Cripple Creek Gold Production Corporation in 1974, and immediate rehabilitation began including 300 feet of timbering in the shaft, installation of a high pressure air line, and work on an improved mill lab. By fall, 1978, $38,000 had been spent improving the mine and on February 14, 1979, the Mary Nevin shipped the first underground ore concentrates produced by the District since the Carlton Mill closed in 1962.

The Henry Adney and Old Gold mines adjoin the Mary Nevin.

HIAWATHA: There was some activity here in 1909 and 1910 with lessees shipping ore in 1911. The property continued to ship from 1935 through 1940.

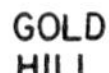

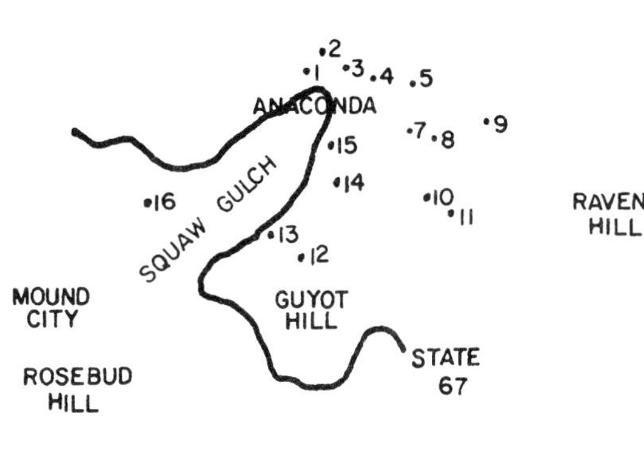

BEACON
HILL

1. ANACONDA
2. VIRGINIA M
3. HOWARD
4. Le CLAIR
5. RAVEN
6. DOLLY VARDEN
7. MORNING GLORY
8. DOCTOR JACK POT
9. INGHAM
10. WEDGE
11. OPHIR
12. KATINKA
13. HOBO
14. ROSE MAUD
15. MARY McKINNEY
16. CALEDONIA

ANACONDA

NOT TO SCALE

ANACONDA

ANACONDA: The area around this mine was consolidated by Denver banker and railroad builder, David H. Moffat, and his mining associate, Eben Smith. The mine survived early stock manipulations, reaching peak production about 1896 from shafts such as the Burke and Fry or Howard, Excelsior, Anaconda Tunnels, and Great View. The mine came under the control of St. Louis businessman John T. Milliken after a voluntary assessment in May, 1900, failed and Milliken engineered the reorganization of the company and assumed the $112,000 debt. The 150-acre property, which had shipped continuously from 1899 to 1903, was then known as the Anaconda Mining Company. Ore shipments continued in 1904 with the Mary McKinney taking over the company.

RAVEL TUNNEL: The Elkton, Gregory, and Tornado all connected with the Anaconda-Raven Tunnel.

MORNING GLORY: It began shipping in 1893 and in 1894 the mine was worked through a tunnel several hundred feet long and numerous surface openings. The Doctor Jack Pot, Ingham, and Morning Glory lodes produced high-grade ore, one four-inch streak of ore worth $40 to $1000 per ton being found on the sixty-foot level in 1896. Sub-lessees, Ronning and Company, found free gold running twenty-five ounces to the ton in the Morning Glory No. 4 in 1901. Lessees continued to work this property in 1906, 1908, and 1911. The Work Company shipped ore from the Morning Glory in 1912, and by 1915 the property was under the control of the Doctor Jack Pot company.

DOCTOR JACK POT: Ore was first discovered in 1893 but when the Doctor Mine was found to be lying on the apex of the Jack Pot vein in 1900, court suits immediately followed. Rather than lose mining profits in litigation, the two mines decided to merge in November, 1901, and the Woods Investment-controlled Doctor Jack Pot was the result. The rich property produced almost continuously until 1945, coming under Carlton interest control around 1928.

INGHAM: Working as early as 1893, it was eventually sold to a Belgian firm with the help of Verner Z. Reed. President of the company in 1900 was Baron William del Marmol of the Brussels, Belgium, branch office of the Ingham Consolidated Gold Mining Company, Limited. In late 1901, the Ingham became the property of the Doctor Jack Pot.

WEDGE: The Wedge precariously displays one of the highest gallows frames on Raven Hill. The mine is best seen from the Anaconda townsite.

OPHIR: In 1983, the Cripple Creek and Victor Mining Company started hauling the 200,000 ton dump of the Ophir Gold Mines Company from the site on Raven Hill. Ore containing .8 ounces of gold per ton was expected.

KATINKA: Also known as the Chicken Hawk, the Katinka Gold Mining Company owned the Katinka, August Flower, Chicken Hawk, and Hobo, about twenty-six acres in all. The Hobo lode was located February 20, 1891, by George Carr and was a prominent mine by 1894.

MARY McKINNEY: The ore body here was located May 27, 1891, by John and Tom Houghton and Frank Castello. Tom and John were experienced prospectors while Frank Castello had a trading post in Florissant, Colorado. The claim was named after the wife of one of the Houghtons. Eventually totaling about 144 acres, the mine shipped almost continuously from 1893 to 1953.

The mine, which was unwatered by the El Paso Drainage Tunnel in 1903, controlled the Anaconda by 1905. By 1912, the Roosevelt Tunnel had drained almost all of the mine, and in 1929 the property was reorganized as the LeClair Consolidated Mining Company. Production stood at almost $6,000,000 by 1932. The last major production came from the huge Mary McKinney dump in 1952 and 1953.

The Anaconda area was also home for the Virginia M., Howard, and LeClair, all of which became part of the Mary McKinney; the Dolly Varden, where a Ft. Carson soldier was killed by gas in 1968; the Rose Maud, which became part of the Doctor Jack Pot after 1915; and the Caledonia, whose manager mortgaged the property to obtain his wages.

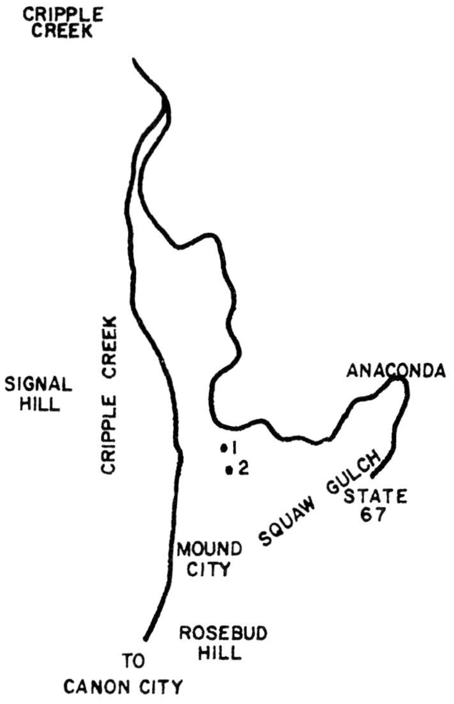

1. GOLD BOND
2. CARDINAL

MOUND CITY
NOT TO SCALE

CARDINAL: Located two hundred feet south of the Caledonia, this operation shipped 1,000 tons of ore containing $20 worth of calaverite per ton. The ore came from a 265-foot incline shaft with two levels and in 1893, a chunk of wire gold was found at this mine. Operations here continued through 1942.

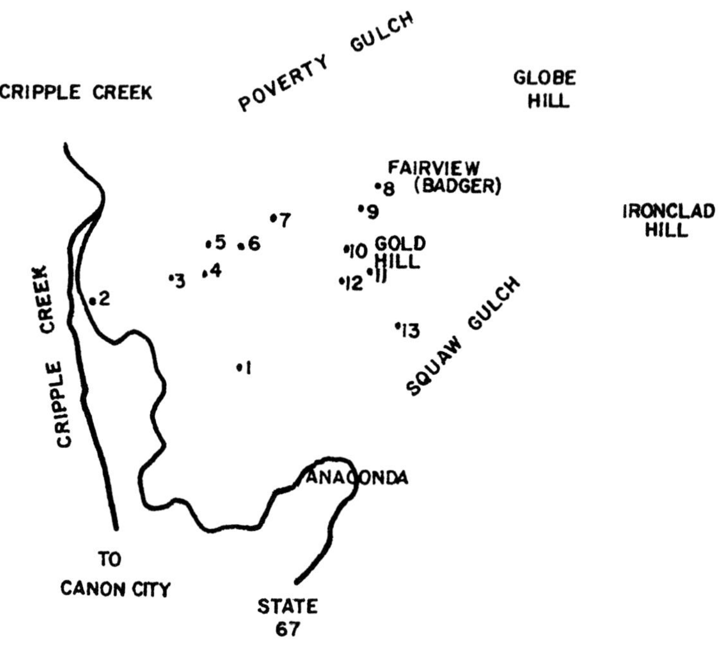

1. INDEX
2. VOLCANO
3. YELLOW BIRD
4. MARIPOSA
5. CONUNDRUM
6. MIDGET
7. MOON-ANCHOR
8. GENEVA
9. JEFFERSON
10. ANCHORIA-LELAND
11. LEXINGTON
12. INTERNATIONAL
13. GRACE GREENWOOD

GOLD HILL
NOT TO SCALE

ANCHORIA-LELAND: Owned by the Anchoria-Leland Mining and Milling Company in 1892, the property consisted of the Anchor, Anchor No. 2, Conundrum, Midland, Lillian Leland, Chance, City View, and Cottontail claims. Values fell off considerably after 1899, but Charles Howbert reopened the mine after 1903 following the labor troubles which had closed it. The mine continued to ship through 1919, two men being killed by gas in the mine in 1911. Ore was shipped in 1925 and 1926 with the Verner Z. Reed Estate property shipping dump ore in 1928.

1931 production was followed by the purchase of the group by the Golden Cycle Corporation in 1935. The mine was acquired for 100,000 shares of the United Gold Mines Company stock and $18,514.95 in cash. Production continued into the 1940's with operations ending on May 1, 1948, due to a shortage of lessees. Total production reached $3,000,000.

The Gold Hill area was one of the first areas to be developed in the District and was the location of the Volcano, Yellow Bird, and the gassy Conundrum, which lost its surface buildings to fire in 1978. Here, too, were the Moon-Anchor, owned at one time by the English firm, Moon-Anchor Consolidated Gold Mines, Limited and the Geneva, a Stratton Estate property. The Jefferson, the Lexington, the International site, scene of much activity by Gold Ore, Limited, and the Grace Greenwood, which produced ore as late as 1959, were also on Gold Hill. (Local History Collection of Pikes Peak Library)

INDEX: Working in 1894, the Index produced from 1908 through 1911 with the El Paso Extension Gold Mines Corporation shipping a considerable amount of ore between 1919 and 1921. A few loads of ore were shipped from the Index group in 1922 when production stood at $923,000. Two men were killed by gas in March, 1925, while repairing a signal line.

Shipments continued from 1926 through 1930 with the Keystone and Lucky Corner lodes producing for Index Mines, (Ltd.) in 1927, and the Fanny Fern Mines Co. which worked the property in 1930. Work continued with the Golden Cycle Corporation purchasing the mine in 1939. The mine and dump were worked sporadically until 1948.

MIDGET: The Midget Gold Mining and Milling Company produced $758,875.69 worth of ore from 1900 to 1904 and ore continued to be shipped through 1909. The mine was only moderately troubled by gas but one miner was killed in 1905 and two in 1919. The Midget-Bonanza King Mill was struck by lightning and destroyed in 1910 after only a few days of operation. The Midget and Bonanza King continued to work from 1911 to 1916, in 1918, in 1923, when the shaft house burned, and from 1925 through 1929. In 1931, the Atlas Gold Mining Company began shipping from the property with ore being produced until 1942. During this period the mine became known as the Atlas. Plans were made to renovate the old mine in the spring of 1981. (The mine in front of the huge Midget dump is the Mariposa.)

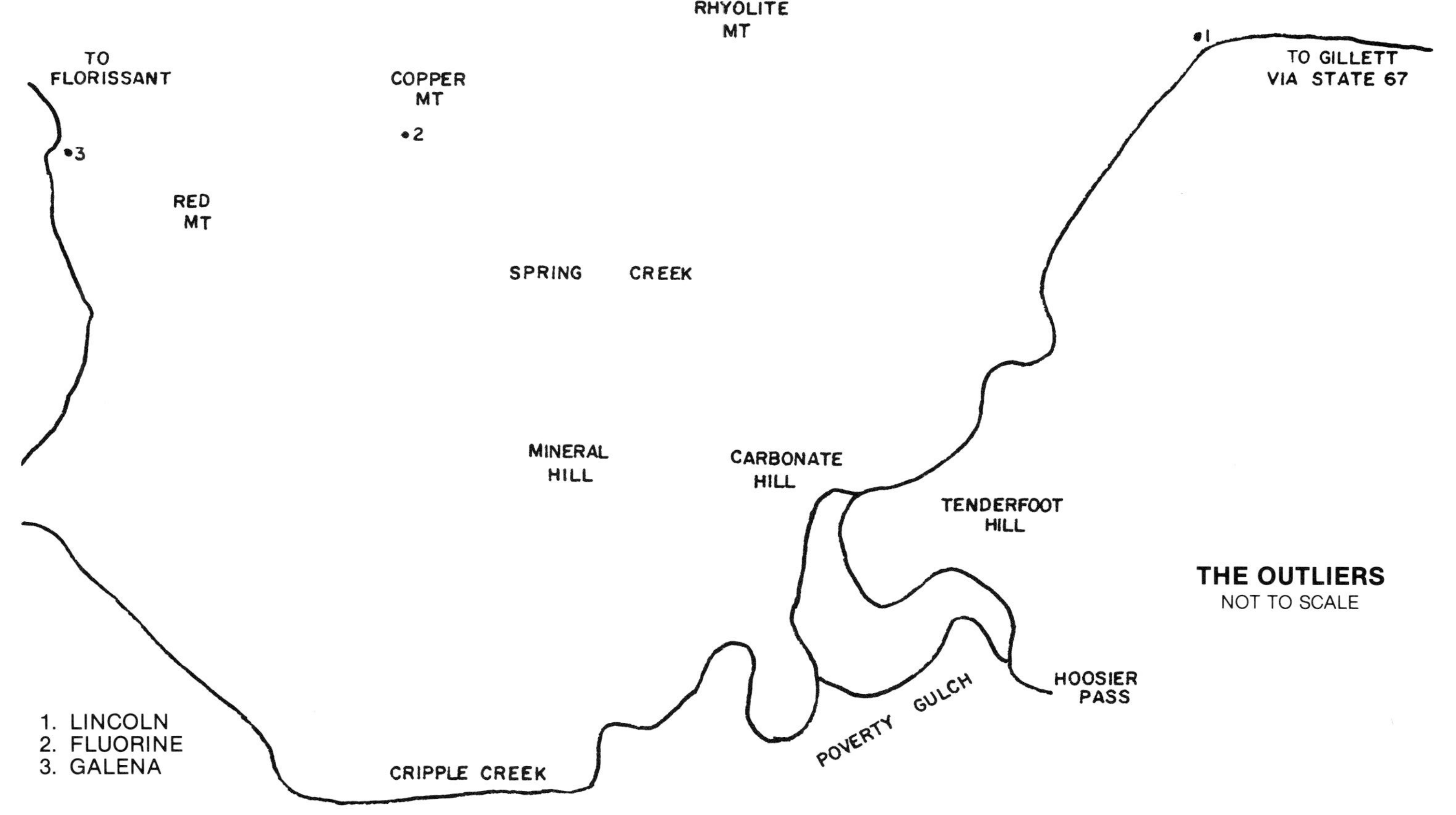

Mines located outside the accepted productive area of Cripple Creek but which produced ore nonetheless were the Lincoln, the Fluorine, located on Copper Mountain, and the Galena Tunnel, which extracted gold and silver from galena ore.

Author's Notes

A wealth of information exists concerning the declining years of the Cripple Creek Mining District after 1900. Government reports, company records, and newspaper articles more than adequately cover this period for the interested researcher. Much time and effort was expended to verify and cross-verify the information and locations presented here.

It will be readily evident to the reader that this book covers the period from the early discovery and boom days to the closing of the Carlton Mill in 1961 but excludes current mining activities in the District. To be sure, Cripple Creek gold is still immensely valuable, with today's prices far exceeding anything imagined by the old hard rock miners. By the same token, state-of-the-art leaching operations of surface dumps somehow lack the romance associated with the Cresson vug or an El Paso Jewelry Shop stope discovery of the old underground mining operations.

With the advent of $800 per ounce gold in 1980, many viable mining operations were organized and exist today in the Cripple Creek District. Complex merger and reorganization programs make it difficult, at best, to trace ownership or operations of the old mines into this period. Although the pursuit of gold ore continues in the District today, the author must be forgiven for his reluctance to drift on these most recently discovered lodes.

Since 1981, Cripple Creek has been the scene of increased mining and milling activity. Obviously, economic instability and general world unrest have been the catalysts which have again awakened the sleeping hills to the sounds of mining operations. Hopefully, this new surge in activity will usher in a whole new era for this old Rocky Mountain gold camp. Certainly no legend deserves it more.

About the Author

Bill Munn, a native of Oklahoma, received his B.S. degree from Phillips University after a stint in Viet Nam where he served as an army artillery officer. First attracted to the Rocky Mountains after several summer vacations from the Great Plains, he returned to the Colorado Springs area in 1973 just as gold fever began to run rampant in many Rocky Mountain gold camps, including Cripple Creek. While living in the Pikes Peak region, he acquired many contacts, both personal and professional, and soon was thoroughly immersed in the Cripple Creek mystique. During the next four years he was able to extensively research the vast amount of information available in the Pikes Peak region and, after his return to Oklahoma, continued to correspond with his contacts in Colorado Springs and to spend his vacations gathering essential information. He now resides with his family in Enid, Oklahoma, where he works as a well-site geologist.